The Rig[...]

by Ma[...]

Table of Contents

Consultant:
Adria F. Klein, Ph.D.
California State University, San Bernardino

capstone
classroom
Heinemann Raintree • Red Brick Learning
division of Capstone

Tens and Ones

Here are 10 chicks. The **numeral** 10 has two **digits**. The first digit, the 1, is in the tens place. The second digit, the 0, is in the ones place.

The 1 stands for 1 group of ten. That is why it is written in the tens place. The 0 stands for no ones. That is why it is written in the ones place.

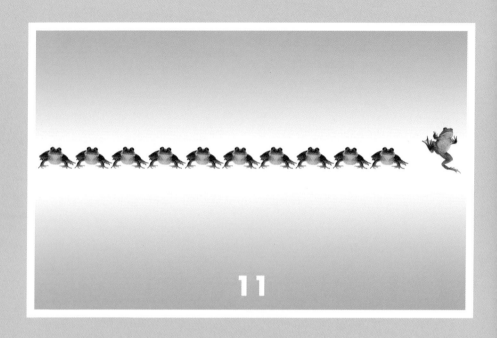

11

Here are 10 frogs. If you add 1 frog to
this group, there will be 11 frogs. The
numeral 11 is written to show that
there is 1 group of ten and 1 one.

Here are 16 ducklings. There is 1 group of ten ducklings. There are 6 other ducklings, or 6 ones. The numeral 1 is written in the tens place. The numeral 6 is written in the ones place. The **total** is 16.

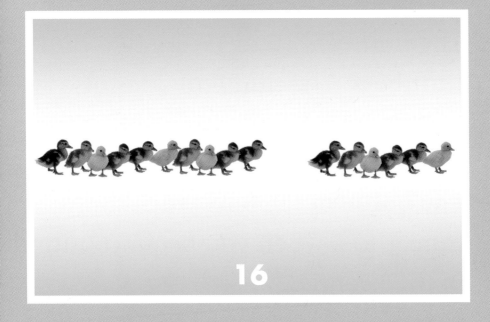

16

More Tens

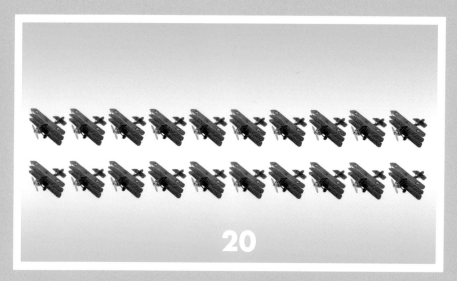

20

The numbers 10 to 19 are **two-digit numbers**. The numeral 20 is also written with two digits. It takes 2 groups of ten and 0 ones to total 20. The **value** of the tens place changes from 1 to 2 when you begin counting in the twenties.

In the picture below, there are 2 groups of ten planes, or 2 tens. There are 7 trucks, or 7 ones. That is a total of 27 vehicles. The 2 is written in the tens place and the 7 is written in the ones place.

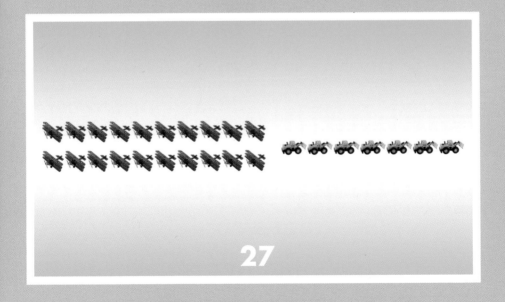

27

In the Right Place

Here is 1 group of ten flowers and 4 more flowers, or 4 ones. The numeral 14 tells how many flowers there are. If the digits were switched, with the 4 in the tens place and the 1 in the ones place, that would say 41 flowers. There aren't that many in the picture!

There are 19 flowers below. There is 1 group of ten flowers plus 9 more, or 9 ones. If the digits were switched, with the 9 in the tens place and the 1 in the ones place, that would say 91 flowers. That's not correct. That's why it is important to put each digit in the right place.

Count the number of kids. There are 12, not 21. There is 1 group of 6 kids plus 6 more kids.

In the chart below, there are 4 groups of ten basketballs plus 1 more ball. That means there are 40 + 1 balls, or 41 balls.

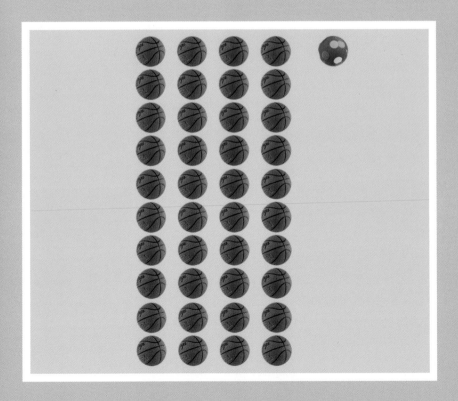

Hundreds Place

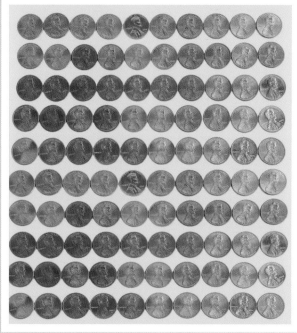

100

One hundred is a **three-digit number**. There is 1 in the hundreds place, 0 tens, and 0 ones. There are 100 pennies on this page.

There are 115 pennies below. That means there is 1 group of a hundred, 1 group of ten, and 5 ones. It takes 10 tens, or 10 groups of ten, to equal 100.

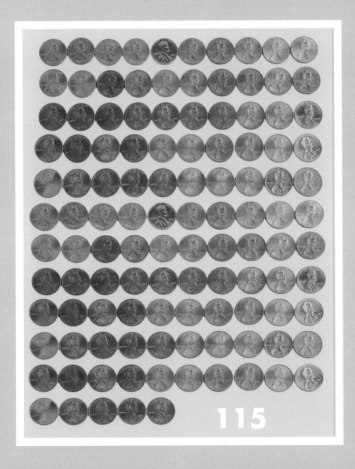

115

The Value of Place

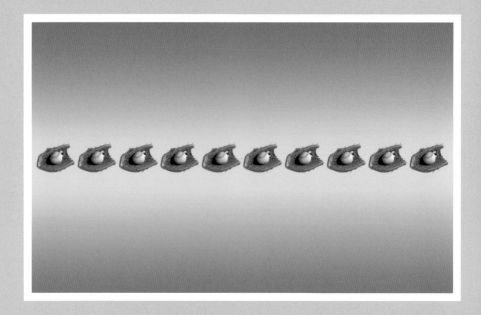

There are 10 gloves in this picture.
There is 1 group of ten gloves. There
are 0 gloves more, or 0 ones. But
you already knew that. You know all
about place value now. Good for you!

Glossary

digit   any numeral from 0 to 9

numeral   the written symbol for a number

place value   the value of the place that each digit has in a numeral

three-digit number   a number with numerals in the hundreds, tens, and ones places

total   the entire amount

two-digit number   a number with numerals in the tens and ones places

value   the amount something is worth

Index